LEVEL C

# NUMBER WORLDS
*Accelerate Math Success*

## Student Workbook

**Sharon Griffin**
**Douglas H. Clements**
**Julie Sarama**

**Authors**

**Sharon Griffin**
*Professor Emerita of Education and Psychology*
Clark University
Worcester, Massachusetts

**Douglas H. Clements**
*Kennedy Endowed Chair in Early
Childhood Learning and Professor*
University of Denver
Denver, Colorado

**Julie Sarama**
*Kennedy Endowed Chair in Innovative
Learning Technologies and Professor*
University of Denver
Denver, Colorado

www.mheonline.com

Copyright © 2015 McGraw-Hill Education

All rights reserved. No part of this publication may be reproduced or distributed in any form or by any means, or stored in a database or retrieval system, without the prior written consent of McGraw-Hill Education, including, but not limited to, network storage or transmission, or broadcast for distance learning.

Send all inquiries to:
McGraw-Hill Education
8787 Orion Place
Columbus, OH 43240

ISBN: 978-0-02-133871-9
MHID: 0-02-133871-X

Printed in the United States of America.

2 3 4 5 6 7 8 9 BRP 28 27 26 25 24

# Contents

| | | |
|---|---|---|
| **Week 1** | Counting | 5 |
| **Week 2** | Counting and Comparing | 8 |
| **Week 3** | More Counting and Comparing | 11 |
| **Week 4** | Matching Dot Sets to Numerals | 14 |
| **Week 5** | Number Sequence and Number Lines | 17 |
| **Week 6** | More Number Sequence and Number Lines | 20 |
| **Week 7** | Number Neighborhoods | 23 |
| **Week 8** | More Number Neighborhoods | 26 |
| **Week 9** | Adding Numbers | 29 |
| **Week 10** | More Adding | 32 |
| **Week 11** | Sequencing Numbers | 35 |
| **Week 12** | Writing Equations | 38 |
| **Week 13** | Counting and Adding | 41 |
| **Week 14** | Making Equations | 44 |
| **Week 15** | Graphing and Comparing Numbers | 47 |
| **Week 16** | More Counting and Adding | 50 |
| **Week 17** | Solving Equations | 53 |
| **Week 18** | Adding and Subtracting | 56 |
| **Week 19** | Subtracting | 59 |
| **Week 20** | Subtracting and Predicting | 62 |
| **Week 21** | Adding and Comparing | 65 |
| **Week 22** | Subtracting to Zero | 68 |
| **Week 23** | More Adding and Subtracting | 71 |

# Contents

**Week 24** Numbers to 100 . . . . . . . . . . . . . . . . . . . . . . . . . . . . . . . . . . . . . . . . . . 74

**Week 25** More Numbers to 100 . . . . . . . . . . . . . . . . . . . . . . . . . . . . . . . . . . . . 77

**Week 26** Addition Stories . . . . . . . . . . . . . . . . . . . . . . . . . . . . . . . . . . . . . . . . . 80

**Week 27** Tens and Ones . . . . . . . . . . . . . . . . . . . . . . . . . . . . . . . . . . . . . . . . . . 83

**Week 28** Adding and Subtracting Length . . . . . . . . . . . . . . . . . . . . . . . . . . . 86

**Week 29** Addition and Subtraction Stories . . . . . . . . . . . . . . . . . . . . . . . . . . 89

**Week 30** Making a Map . . . . . . . . . . . . . . . . . . . . . . . . . . . . . . . . . . . . . . . . . . . 92

**Week 31** Understanding the Analog Clock . . . . . . . . . . . . . . . . . . . . . . . . . . 95

**Week 32** The Ten-Dollar Bill and the Dime . . . . . . . . . . . . . . . . . . . . . . . . . . 98

**Practice Writing** . . . . . . . . . . . . . . . . . . . . . . . . . . . . . . . . . . . . . . . . . . . . . . . . 101

# Week 1 • Counting

## Drop and Count

Name _____  Date _____

**Write** the number story.

1.

   _____ + _____ = _____

2.

   _____ + _____ = _____

3.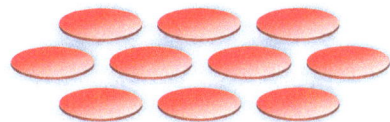

   _____ + _____ = _____

4.

    +  =

# Week 1 • Counting

## Feed the Animals

Name _____ Date _____

**Draw** a line from the Counters to the number that shows how many.

1. 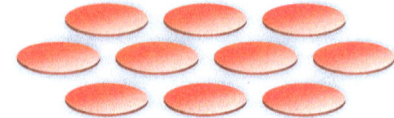     11

2.       14

3.      20

4.       17

# Week 1 • Counting

## Review

Name _____ Date _____

**Write** the missing number.

1. 2, 3, _____, 5, 6

2. 7, _____, 9, 10, 11

3. 12, 13, 14, _____, 16

4. 17, 18, _____, 20, 21

# Week 2 • Counting and Comparing

## Count and Compare

Name _____  Date _____

**Write** how many.

1. 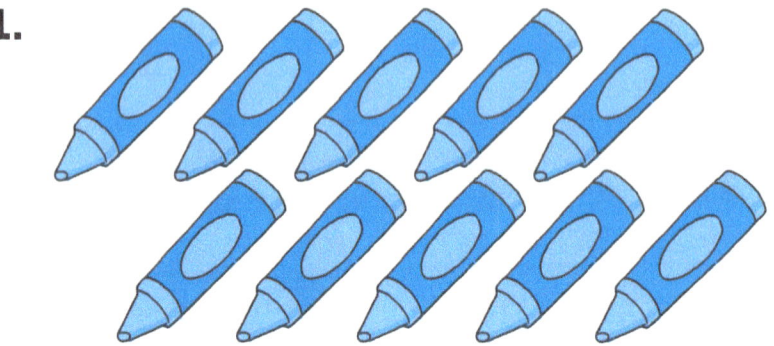 _____

2.  _____

3.    _____

4.   _____

8  Level C

**Week 2 • Counting and Comparing**

# Food Fun

Name _____ Date _____

## Write the answer.

**1.** How many children?

_____

**2.** How many apples?

_____

**3.** Do you need more apples?

_____

**4.** How many more do you need?

_____

# Week 2 • Counting and Comparing

## Review

Name _____  Date _____

**Circle** the group that has more.

1.

2.

3.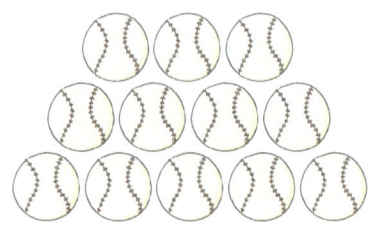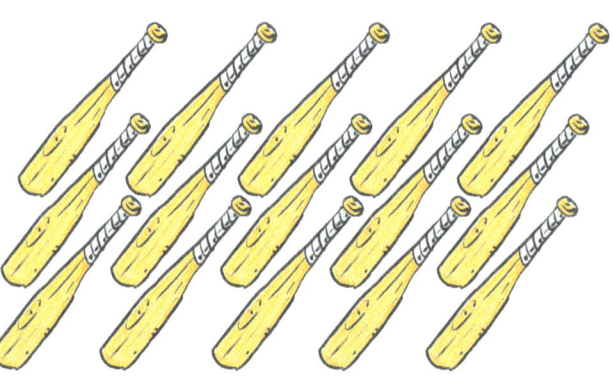

4.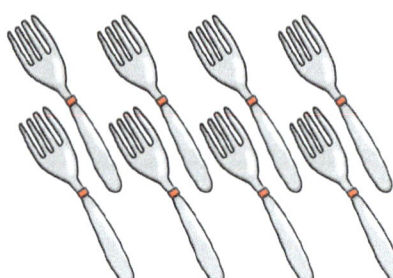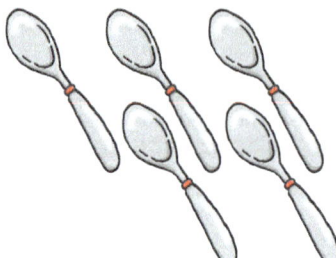

10  Level C

Week 3 • More Counting and Comparing

# Concentration

Name _____  Date _____

**Circle** the matching card.

1.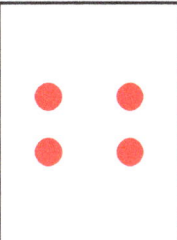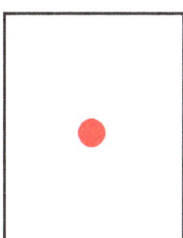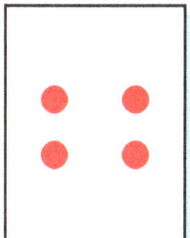

2.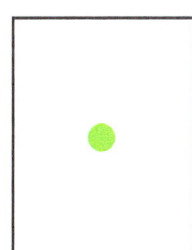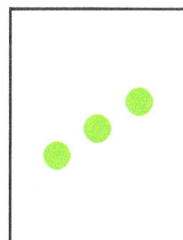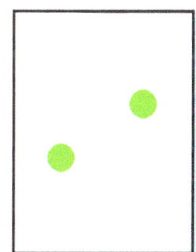

3.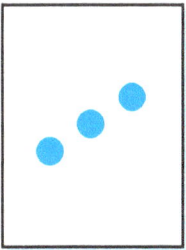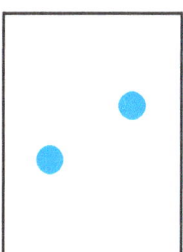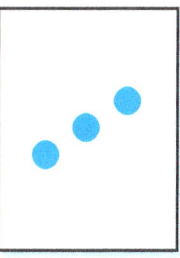

4.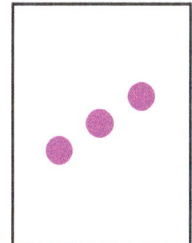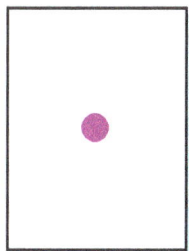

# Week 3 • More Counting and Comparing

## Party!

Name _____  Date _____

## Complete

1.

   Are there more hats or oranges? _____

2.

   Are there enough balloons
   for each child? _____

3.

   Are there too few or
   too many horns? _____

4.

   Are there enough hats? _____

12  Level C

# Week 3 • More Counting and Comparing

# Review

Name _____  Date _____

**Look** at the Dot Set Card.
Circle the matching numeral.

1. 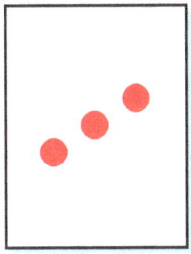   5   3   2

2. 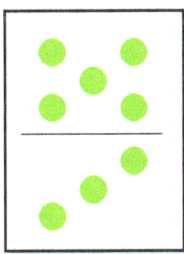   8   6   7

**Circle** the group that has more.

3.

4. How many hats do you need?

   2   4   3

Week 3 **More Counting and Comparing** 13

# Week 4 • Matching Dot Sets to Numerals

## Concentration to 20

Name _____  Date _____

**Look** at the Dot Set Cards.
Circle the matching number.

1.    5   11   7

2. 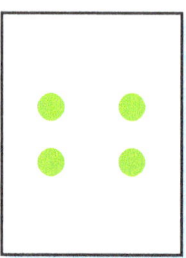   4   5   6

3. 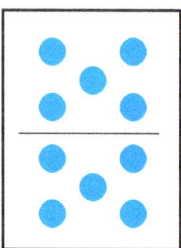 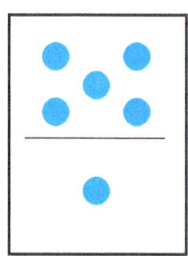   10   6   16

4. 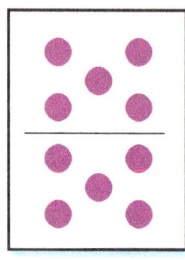 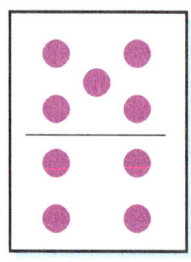   10   19   23

**Week 4 • Matching Dot Sets to Numerals**

# Bravo!

Name _____  Date _____

**Circle** the greater number.

1.  14    12

2.  14    17

3.  19    20

4.  16    13

# Week 4 • Matching Dot Sets to Numerals

## Review

Name _____ Date _____

**Circle** the greater number.

1.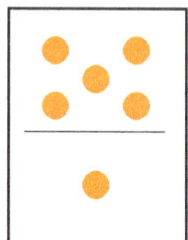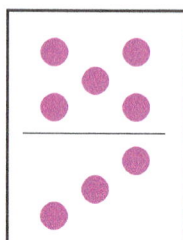

2.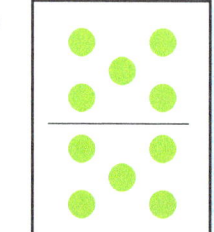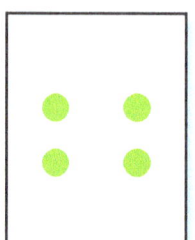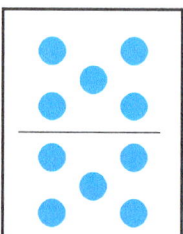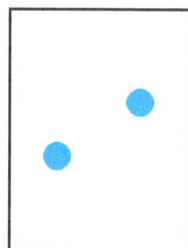

3.

   17         15

4.

   17         19

# Week 5 • Number Sequence and Number Lines

## Delivering Mail

Name _____  Date _____

**Put** the numbers in order from smallest to biggest.

1.  14    6    19    17

    ____  ____  ____  ____

2.  2    15    11    8

    ____  ____  ____  ____

3.  18    13    10    16

    ____  ____  ____  ____

4.  14    9    20    4

    ____  ____  ____  ____

# Week 5 • Number Sequence and Number Lines

# Magnetic Number Line Game

Name _____ Date _____

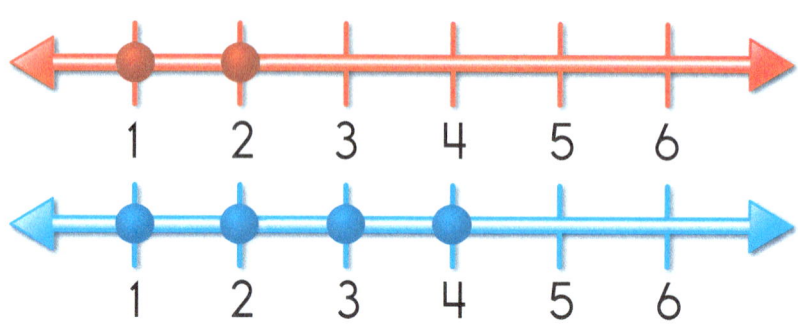

**Answer** each question.

1. Who has the most?

   _____

2. Who has the least?

   _____

3. How many red Counters do you need to be on 5?

   _____

4. How many blue Counters do you need to be on 5?

   _____

18  Level C

Week 5 • Number Sequence and Number Lines

# Review

Name _____  Date _____

**Put** the numbers in order from smallest to biggest.

1. 8   17   12   5

   ____   ____   ____   ____

2. 4   14   6   16

   ____   ____   ____   ____

3. Add 2 dots.

4. Add 3 more dots.

5. How many dots do you have? _____

**Week 6 • More Number Sequence and Number Lines**

# Delivering Mail

Name _____ Date _____

## Answer each question.

**1.** Circle the number that is greater than 14.

    12                15

**2.** Circle the number that comes after 8.

    18                6

**3.** Circle the number that is smaller than 15.

    20              10

**4.** Circle the number that comes before 7.

    6                8

Week 6 • More Number Sequence and Number Lines

# Plus-Minus Game

Name _____  Date _____

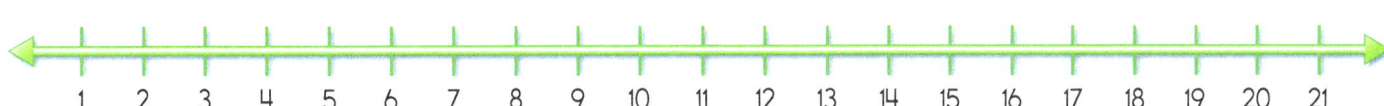

**Write** what number you will be on.

1. 5      _____

2. 8      _____

3. 13     _____

4. 16     _____

# Week 6 • More Number Sequence and Number Lines

## Review

Name _____ Date _____

**Answer** each question.

1. Circle the number that comes before 13.

    12          14

2. Circle the number that comes after 5.

    13          3

3.

    9 − 1 = _____

4.

    16 + 1 = _____

**Week 7** • Number Neighborhoods

# Numbering the Neighborhood

Name _____ Date _____

**Write** the missing number.

1.

_____  2   3   4   5

2.

6   7   8   _____   10

3.

12   13   _____   15   16

4.

16   _____   18   19   20

Week 7 • Number Neighborhoods

# The Biggest Neighborhood

Name _____  Date _____

**Circle** the correct answer

1. You're on 5.
   What card do you need
   to get to 6?

2. You're on 8.
   What card do you need
   to get to 6?

3. You're on 10.
   What card do you need
   to get to 9?

4. You're on 2.
   What card do you need
   to get to 4?

# Week 7 • Number Neighborhoods

# Review

Name _____ Date _____

**Write** the missing number.

1. 2  3  ____  5  6

2. 12  13  14  15  ____

3. 9  ____  11  12  13

4. ____  16  17  18  19

**Week 8 • More Number Neighborhoods**

# Meet Your Neighbors

Name _____ Date _____

**Draw** a line to the next-door neighbor.

1. 8      4

2. 12      11

3. 3      16

4. 15      7

**Week 8 • More Number Neighborhoods**

# Building a Neighborhood

Name _____ Date _____

Try to get three in a row.

**Draw** a line to the number you need.

1. | 6 | 8 |   | 5 |

2. | 18 | 19 |   | 7 |

3. | 13 | 14 |   | 15 |

4. | 3 | 4 |   | 17 |

**Week 8 • More Number Neighborhoods**

# Review

Name _____ Date _____

**Write** the numbers that are just before and just after each number below.

1. | 12 |    _____    _____

2. | 4 |    _____    _____

3. | 19 |    _____    _____

4. | 14 |    _____    _____

**Week 9 • Adding Numbers**

# Can You Guess My Number?

Name _____  Date _____

**Circle** the correct answer.

**1.** It is smaller than 8.

        7             17

**2.** It is smaller than 18.

       17            20

**3.** It is 1 more than 9.

       19            10

**4.** It is a neighbor of 7.

       15            6

# Week 9 • Adding Numbers

## Dragon Quest 1

Name _____   Date _____

**Write** the number story.

1.

   _____  _____ = _____

2.

   _____  _____ = _____

3.

   _____  _____ = _____

4.

   _____  _____ = _____

# Week 9 • Adding Numbers

# Review

Name _____  Date _____

## Circle the correct answer.

**1.** It is a neighbor of 20.

      19          15

**2.** It is bigger than 8.

      6          14

## Write the number story.

**3.**     

_____ + _____ = _____

**4.**     

_____ + _____ = _____

**Week 10 • More Adding**

# Can You Guess My Number?

Name _____ Date _____

**Circle** the correct answer.

1. It is 1 more than 14.

    15        13

2. It is smaller than 15.

    9        19

3. It is a neighbor of 6.

    2        7

4. It is 1 less than 18.

    17        19

# Week 10 • More Adding

# Drop and Add

Name _____ Date _____

**Write** the number story.

1.

   _____ + _____ = _____

2.

   _____ + _____ = _____

3.

   _____ + _____ = _____

4.

   _____ + _____ = _____

# Week 10 • More Adding

## Review

Name _____ Date _____

**Circle** the correct answer.

**1.** It is a neighbor of 11.

      10          20

**2.** It is 1 more than 3.

      4          14

**Write** the number story.

**3.**

_____  + _____ = _____

**4.**

_____  + _____ = _____

34 Level C

**Week 11** • Sequencing Numbers

# And One More Makes . . .

Name _____ Date _____

**Write** the answer.

1. 
$1 + 1 = $ _____

2. 
$1 + 1 + 1 = $ _____

3. 
$1 + 1 + 1 + 1 = $ _____

4. 
$1 + 1 + 1 + 1 + 1 = $ _____

**Week 11 • Sequencing Numbers**

# More Food Fun

Name _____   Date _____

**Write** the missing number.

1. 🔵🔵🔵🔵🔵  +  =  🔵🔵🔵🔵🔵🔵

   5  +  0  =  _____

2. 🟢  +  🟢🟢🟢🟢  =  🟢🟢🟢🟢🟢

   _____  +  4  =  5

3. 🔴🔴  +  🔴🔴🔴  =  🔴🔴🔴🔴🔴

   2  +  _____  =  5

4. +  🟡🟡🟡🟡🟡  =  🟡🟡🟡🟡🟡

   _____  +  5  =  5

Week 11 • Sequencing Numbers

# Review

Name _____ Date _____

**Write** the answer.

1. 0 + 1 = _____

2. 1 + 1 = _____

3. 2 + 0 = _____

4. 2 + 1 = _____

# Week 12 • Writing Equations

## They're Gone

Name _____  Date _____

**Write** the number story

1. 🟡🟡🟡🟡🟡  —  🟡🟡🟡🟡  =  🟡

   _____  —  _____  =  _____

2. 🔴🔴🔴🔴🔴  —  🔴🔴  =  🔴🔴🔴

   _____  —  _____  =  _____

3. 🟢🟢🟢🟢🟢  —     =  🟢🟢🟢🟢🟢

   _____  —  _____  =  _____

4. 🔵🔵🔵🔵🔵  —  🔵  =  🔵🔵🔵🔵

   _____  —  _____  =  _____

38  Level C

**Week 12 • Writing Equations**

# Add Them Up

Name _____  Date _____

**Write** the number story

1. 2 + 1 + 3 = 6

   ___ + ___ + ___ = ___

2. 1 + 1 + 2 = 4

   ___ + ___ + ___ = ___

3. 3 + 1 + 3 = 7

   ___ + ___ + ___ = ___

4. 5 + 2 + 3 = 10

   ___ + ___ + ___ = ___

# Week 12 • Writing Equations

## Review

Name _____  Date _____

**Write** the number story

1. 🟢🟢🟢🟢 − 🟢🟢🟢 = 🟢

   _____

2. 🟠🟠 + 🟠🟠🟠 + 🟠🟠🟠🟠 = 🟠🟠🟠🟠🟠🟠🟠🟠🟠

   _____

3. 🔵🔵🔵🔵🔵 + 🔵 + 🔵 = 🔵🔵🔵🔵🔵🔵🔵

   _____

4. 🔴🔴🔴🔴🔴 − 🔴🔴🔴 = 🔴🔴

   _____

# Week 13 • Counting and Adding

## Steve's New Bike

Name _____  Date _____

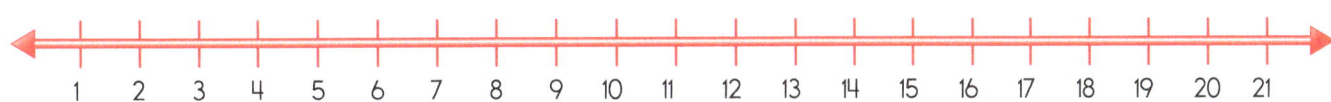

**1.** Circle the number that is closer to 0.

        12        21

**2.** Circle the number that is farther from 0.

        19        9

**3.** Circle the number that is closer to 100.

        69        71

**4.** Circle the number that is farther from 100.

        49        64

# Week 13 • Counting and Adding

## Secret Number Game

Name _____ Date _____

**Write** the answer.

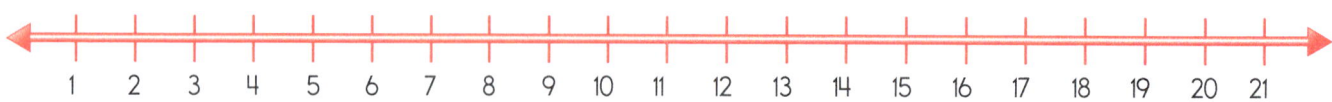

1. 

    0 + 3 = _____

2. 

    6 + 4 = _____

3. 

    11 + 5 = _____

4. 

    9 + 6 = _____

# Week 13 • Counting and Adding

## Review

Name _____ Date _____

**Write** the answer.

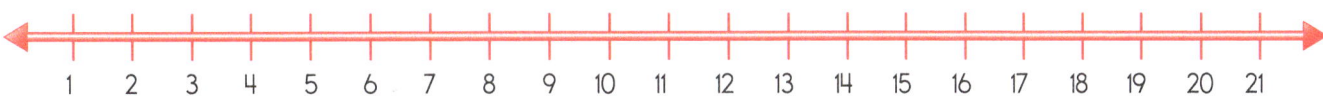

**1.**

10 + 4 = _____

**2.**

4 + 6 = _____

**3.**

12 + 3 = _____

**4.**

8 + 5 = _____

Week 14 • Making Equations

# Making Equations for Beginners

Name _____  Date _____

**Draw** a line from each group of Dot Set Cards to show how much the cards equal.

1. 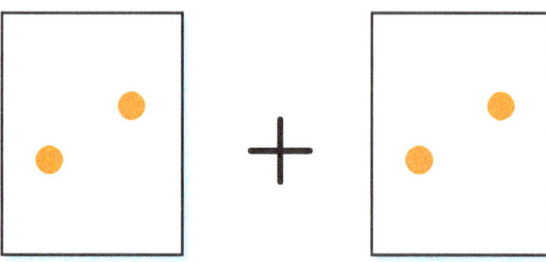 5

2. 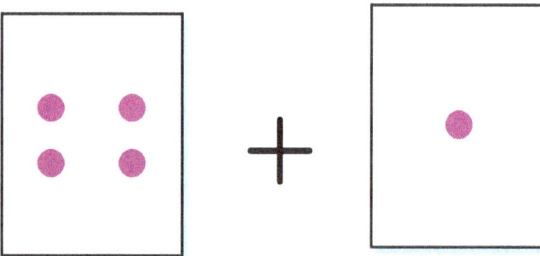 2

3. 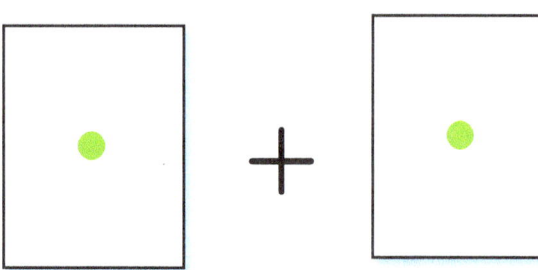 4

4. 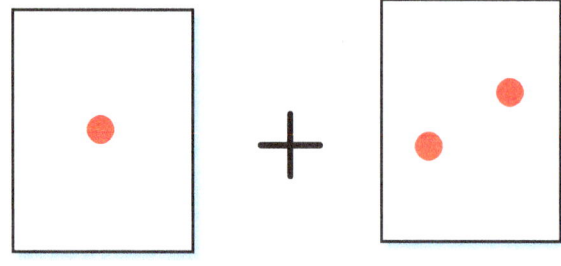 3

# Week 14 • Making Equations

## Making and Writing Equations

Name _____  Date _____

**Write** the number story.

1. 🟢🟢 + 🟢🟢🟢🟢🟢 = 🟢🟢🟢🟢🟢🟢🟢

   _____

2. 🟠🟠🟠🟠🟠🟠 + 🟠🟠🟠 = 🟠🟠🟠🟠🟠🟠🟠🟠🟠

   _____

3. 🔵🔵🔵 + 🔵🔵 = 🔵🔵🔵🔵🔵

   _____

4. 🔴🔴🔴🔴 + 🔴 = 🔴🔴🔴🔴🔴

   _____

Week 14 **Making Equations** 45

# Week 14 • Making Equations

## Review

Name _____  Date _____

**Write** the number story.

1. 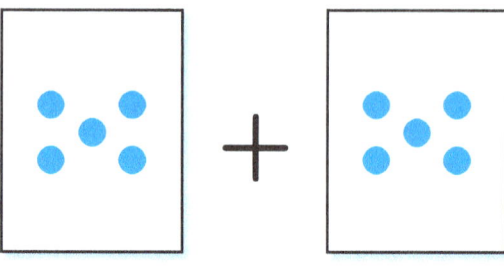 _____

2. 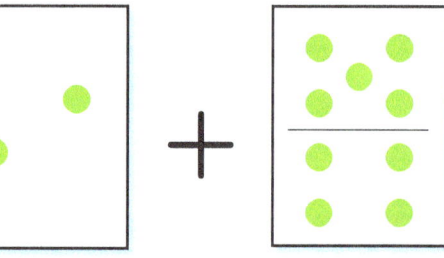 _____

3. 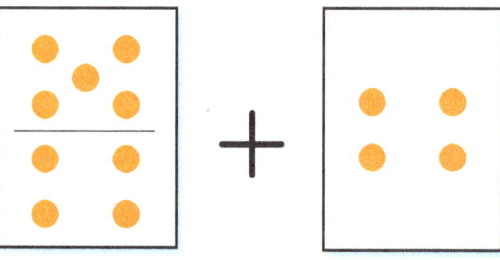 _____

4. 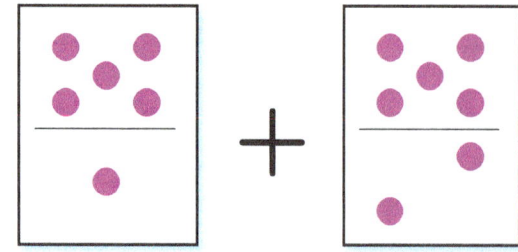 _____

# Week 15 • Graphing and Comparing Numbers

## Linking and Thinking

Name _____  Date _____

### Write how many.

1.   How many? _____

2.   How many? _____

3.   How many? _____

### Graph how many.

4.

# Week 15 • Graphing and Comparing Numbers

## Making and Writing Equations

Name _____  Date _____

**Write** the number story.

1.

   _____

2.

   _____

3.

   _____

4.

   _____

# Week 15 • Graphing and Comparing Numbers

# Review

Name _____ Date _____

## Write how many.

1. 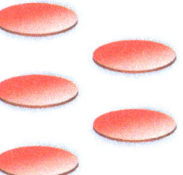   How many? _____

2.

      How many? _____

3.
      How many? _____

## Graph how many.

4.

# Week 16 • More Counting and Adding

## Secret Shopping Game

Name _____ Date _____

**Write** the number story.

**1.**

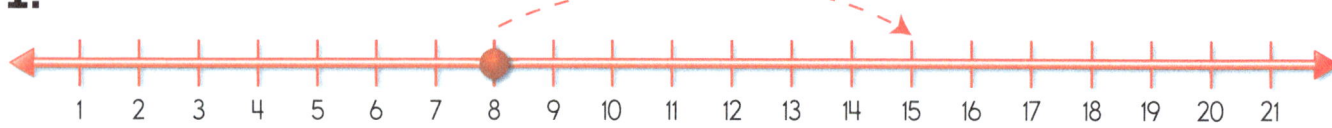

_____

**2.**

_____

**3.**

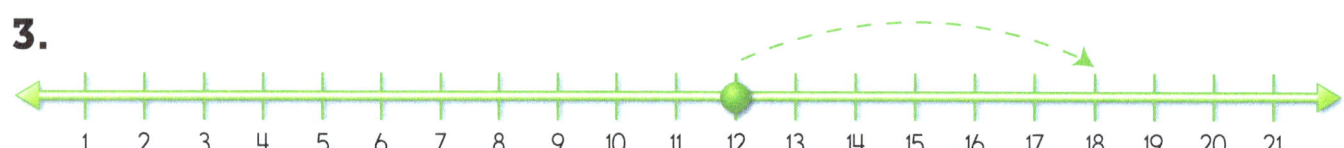

_____

**4.**

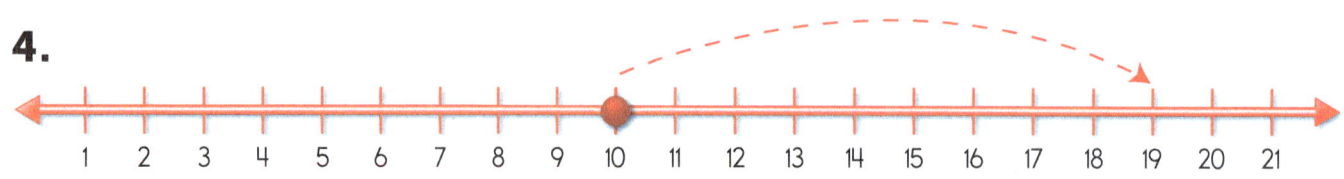

_____

**Week 16** • More Counting and Adding

# Changing Addends

Name _____  Date _____

**Write** the number story.

1.

   _____

2.

   _____

3.

   _____

4.

   _____

# Week 16 • More Counting and Adding

## Review

Name _____ Date _____

**Write** the number story.

**1.**
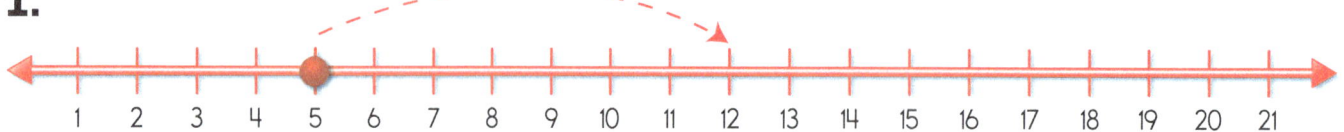
_____

**2.**
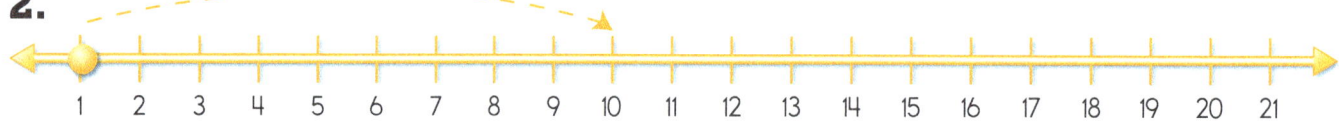
_____

**3.**
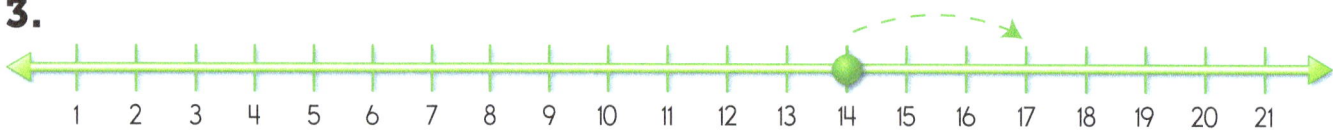
_____

**4.**

_____

# Week 17 • Solving Equations

# Three Area Counter Drop

Name _____ Date _____

**Write** the number story.

1.

   _____

2.

   _____

3.

   _____

4.

   _____

# Week 17 • Solving Equations

# Counter Dropping

Name _____ Date _____

**Write** the number story.

1.

   _____

2.

   _____

3.

   _____

4.

   _____

# Week 17 • Solving Equations

## Review

Name _____ Date _____

**Write** the number story.

1.

   _____

2.

   _____

3.

   _____

4.

   _____

# Week 18 • Adding and Subtracting

## Dragon Quest 2

Name _____ Date _____

**Write** the answer.

1. 3 − 2 = _____

2. 3 + 2 = _____

3. 4 + 5 = _____

4. 4 − 2 = _____

# Week 18 • Adding and Subtracting

## Shopping Trip

Name _____ Date _____

**Write** the number story.

1.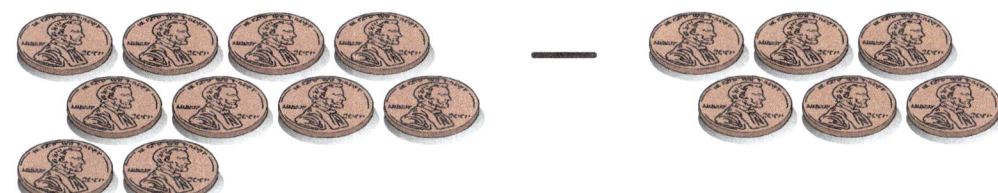

   _____

2.  —

   _____

3.  —

   _____

4.  —

   _____

# Week 18 • Adding and Subtracting

## Review

Name _____ Date _____

**Write** the answer.

1. $9 - 5 = \underline{\phantom{00}}$

2. $7 - 4 = \underline{\phantom{00}}$

3. $6 - 5 = \underline{\phantom{00}}$

4. $2 - 2 = \underline{\phantom{00}}$

# Week 19 • Subtracting

## Tear It Up and Throw It Away

Name _____ Date _____

**Write** the answer.

1. 18 − 6 = _____

2. 10 − 4 = _____

3. 15 − 3 = _____

4. 8 − 5 = _____

# Week 19 • Subtracting

## Subtracting Counters

Name _____ Date _____

**Write** the number story.

1.

   _____

2.

   _____

3.

   _____

4.

   _____

# Week 19 • Subtracting

## Review

Name _____  Date _____

**Write** the number story.

**1.**

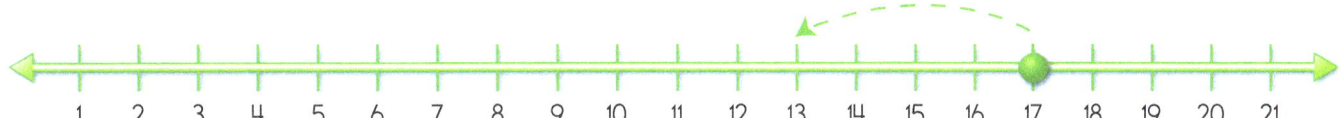

_____

**2.**

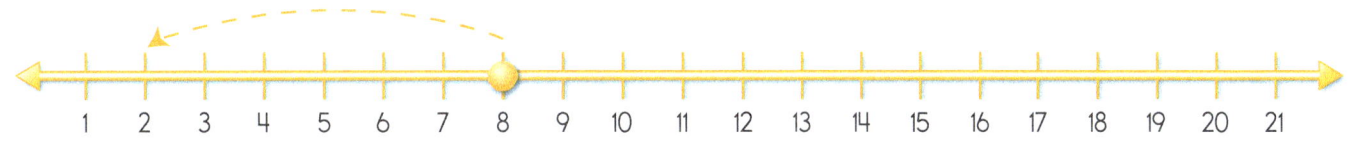

_____

**3.**

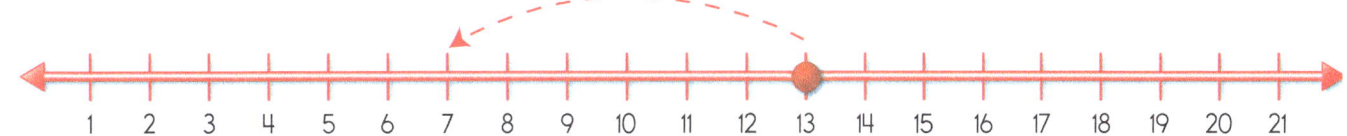

_____

**4.**

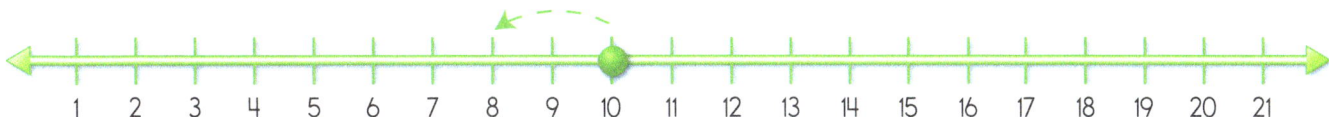

_____

# Week 20 • Subtracting and Predicting

## Going Fishing

Name _____  Date _____

**Write** the answer.

1. 14 − 7 = _____

2. 8 − 2 = _____

3. 11 − 5 = _____

4. 5 − 1 = _____

**Week 20 • Subtracting and Predicting**

# Adding Concentration

Name _____  Date _____

**Write** the missing number.

1. 
$3 + \underline{\phantom{xx}} = 5$

2. 
$\underline{\phantom{xx}} + 1 = 5$

3. 
$2 + \underline{\phantom{xx}} = 5$

4. 
$5 + \underline{\phantom{xx}} = 5$

# Week 20 • Subtracting and Predicting

## Review

Name _____  Date _____

**Write** the missing number.

1. 10 − 8 = _____

2. 7 − 2 = _____

3. 4 + _____ = 5

4. 5 + 0 = _____

# Week 21 • Adding and Comparing

## Flip, Add, and Compare

Name _____  Date _____

**Solve** each problem.
**Circle** the bigger sum.

**1.**

10 + 3 = _____

12 + 0 = _____

**2.**

7 + 4 = _____

8 + 4 = _____

# Week 21 • Adding and Comparing

## Roll, Add, and Compare

Name _____  Date _____

**Solve** each problem.
**Circle** the bigger sum.

**1.**

$5 + 4 =$ _____

$4 + 3 =$ _____

**2.**

$2 + 2 =$ _____

$3 + 3 =$ _____

# Week 21 • Adding and Comparing

## Review

Name _____ Date _____

**Solve** each problem.
**Circle** the bigger sum.

**1.**

14 + 4 = _____

8 + 9 = _____

**2.**

6 + 8 = _____

12 + 4 = _____

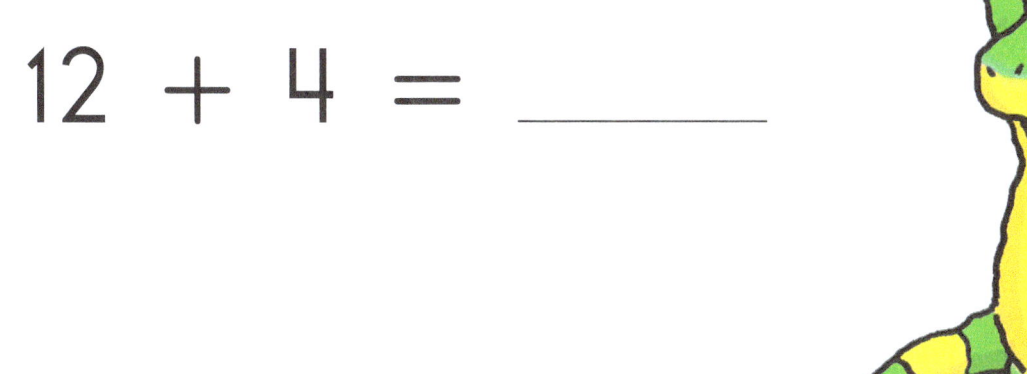

**Week 22 • Subtracting to Zero**

# Subtracting to Zero

Name _____  Date _____

**Write** the answer.

1. 
    12 − 12 = _____

2. 
    7 − _____ = 0

3. 
    4 − 4 = _____

4. 
    15 − _____ = 0

**Week 22 • Subtracting to Zero**

# Which Way Should I Go?

Name _____ Date _____

**Solve** each problem.
**Circle** the answer closer to 0.

**1.**

9 + 3 = _____

15 − 2 = _____

**2.**

7 − 3 = _____

2 + 3 = _____

# Week 22 • Subtracting to Zero

## Review

Name _____ Date _____

**Write** the answer.

1. 13 − 13 = _____

2. 9 − _____ = 0

3. 8 − 8 = _____

4. 1 − _____ = 0

**Week 23** • More Adding and Subtracting

# Counting Back

Name _____  Date _____

**Write** the number story.

1. $\boxed{20} - \boxed{4}$ _____

2. $\boxed{14} - \boxed{3}$ _____

3. $\boxed{9} - \boxed{2}$ _____

4. $\boxed{15} - \boxed{5}$ _____

# Week 23 • More Adding and Subtracting

## Dragon Quest 3

Name _____  Date _____

**Write** the answer.

1. 5 − 2 = _____

2. 2 + 5 = _____

3. 4 + 3 = _____

4. 3 − 1 = _____

# Week 23 • More Adding and Subtracting

## Review

Name _____  Date _____

**Write** the answer.

1. 13 − 6 = _____

2. 8 − 4 = _____

3. 12 − 3 = _____

4. 10 − 4 = _____

# Week 24 • Numbers to 100

## Elevator Game

Name _____  Date _____

**Write** the answer.

1. 
   8 − 4 = _____

2. 
   10 − 3 = _____

3. 
   4 − 2 = _____

4. 
   7 − 3 = _____

**Week 24 • Numbers to 100**

# Number Line to 100

Name _____  Date _____

**Write** the missing number.

**1.**

43  44  _____  46  47

**2.**

24  _____  26  27  28

**3.**

68  69  _____  71  72

**4.**

96  97  98  _____  100

# Week 24 • Numbers to 100

## Review

Name _____ Date _____

**Write** the answer.

1. 

    5 − 4 = _____

2. 

    14 − 2 = _____

**Write** the missing number.

3. 

    28  _____  30  31  32

4. 

    55  56  57  _____  59

76 Level C

# Week 25 • More Numbers to 100

## Steve's New Bike

Name _____  Date _____

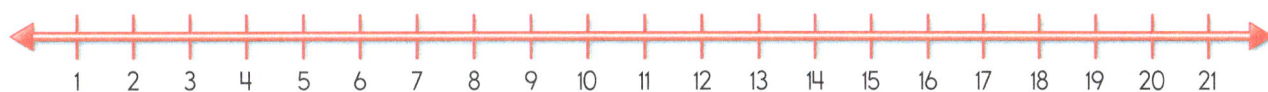

**Circle** the answer.

1. Circle the number that is closer to 0.

   32          23

2. Circle the number that is farther from 0.

   44          54

3. Circle the number that is closer to 100.

   63          37

4. Circle the number that is farther from 100.

   72          78

# Week 25 • More Numbers to 100

## Writing Numbers to 100

Name _____   Date _____

**Write** the missing numbers.

1.

20  _____  _____  23  24

2.

48  _____  50  _____  52

3.

26  27  _____  _____  30

4.

67  68  _____  _____  71

# Week 25 • More Numbers to 100

## Review

Name _____  Date _____

**Write** the missing numbers.

1. 18  19  _____  _____  22

2. 52  _____  _____  55  56

3. 60  61  _____  _____  64

4. 88  _____  90  _____  92

# Week 26 • Addition Stories

## How Many Balloons?

Name _____  Date _____

**Write** and solve the word problems.

**1.**

I had 5 fish.

Mom gave me 3.

How many fish do I have now?

_____

**2.**

I ate one apple.

My friend ate 3.

How many apples did we eat?

_____

# Week 26 • Addition Stories

## And the Answer Is …

Name _____  Date _____

**Match** the Dot Set Cards to the Number Cards.

1. 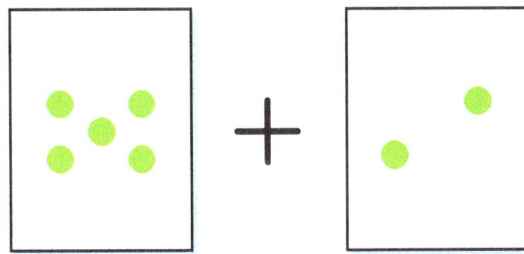   10

2. 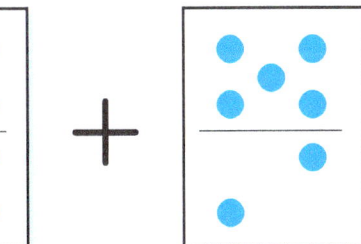   14

3. 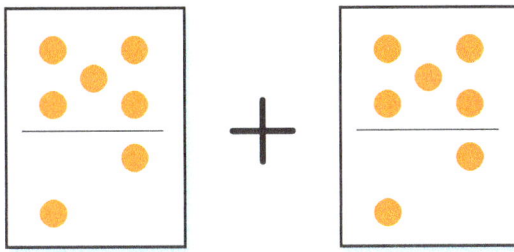   7

4. 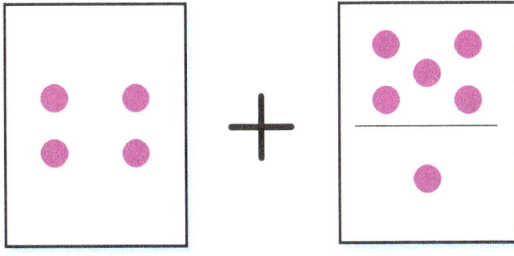   17

# Week 26 • Addition Stories

## Review

Name _____  Date _____

**Write** and solve the word problems.

**1.**

I had 4 crayons.

I got 5 more.

How many crayons do I have now?

_____

**2.**

I had 3 nuts.

Dad gave me 2 more.

How many nuts do I have now?

_____

# Week 27 • Tens and Ones

# Rosemary's Super Shoes

Name _____ Date _____

**Write** the answer.

1. **47**

   How many tens? _____

   How many ones? _____

2. **26**

   How many tens? _____

   How many ones? _____

3. **60**

   How many tens? _____

   How many ones? _____

4. **35**

   How many tens? _____

   How many ones? _____

# Week 27 • Tens and Ones

## Dragon Quest 4

Name _____  Date _____

**Write** the answer.

1. 7 + 9 = _____

2. 14 − 4 = _____

3. 12 + 8 = _____

4. 13 + 3 = _____

84   Level C

**Week 27 • Tens and Ones**

# Review

Name _____ Date _____

**Write** the answer.

1. 98

   How many tens? _____

   How many ones? _____

2. 53

   How many tens? _____

   How many ones? _____

3. 

   9 + 9 = _____

4. 

   6 + 8 = _____

# Week 28 • Adding and Subtracting Length

## How Much More?

Name _____ Date _____

**Write** the answer.

**1.**

How tall is the big dog? _____

**2.**

How tall is the small dog? _____

**Write** the number story.

**3.** How much taller is the big dog?

_____

**4.** Circle the minus sign.

# Week 28 • Adding and Subtracting Length

## Let's Add Some More

Name _____  Date _____

**Write** the answer.

1. 6 + 8 = _____

2. 12 + 7 = _____

3. 8 + 6 = _____

4. 10 + 5 = _____

# Week 28 • Adding and Subtracting Length

## Review

Name _____  Date _____

**Write** the answer.

**1.**

14 + 6 = _____

**2.**

9 − 7 = _____

**3.**

4 + 12 = _____

**4.**

15 − 4 = _____

# Week 29 • Addition and Subtraction Stories

## How Many Are Left?

Name _____ Date _____

**Write** the number stories.

1. I had 14 flowers. I gave 5 to my friend. How many are left?

   _____

2. Then I gave 3 flowers to my brother. How many are left?

   _____

3. Next I gave 2 flowers to my teacher. How many are left?

   _____

**Week 29 • Addition and Subtraction Stories**

# Going Shopping

Name _____  Date _____

**Write** the answer.

1. 5 + 3 + 2 = _____

2. 4 + 7 + 4 = _____

3. 3 + 9 + 5 = _____

4. 3 + 6 + 3 = _____

# Week 29 • Addition and Subtraction Stories

## Review

Name _____  Date _____

**Write** the answer.

1. 5 + 3 + 4 = _____

2. 6 − 2 − 2 = _____

3. 4 + 5 + 6 = _____

4. 3 + 12 + 4 = _____

**Week 30 • Making a Map**

# Feed the Pets

Name _____  Date _____

**Write** and solve the number stories.

1. My dog eats 2 cups of food each day. How much does he eat in 2 days?

   _____

2. My cat eats 1 cup of food each day. How much does she eat in 2 days?

   _____

3. How much do they both eat in 2 days?

   _____

# Week 30 • Making a Map

## Hallway Map

Name _____  Date _____

**Write** the answer.

**1.**

I pass _____ rooms on the way to Room 5.

**2.**

Room _____ is before Room 2.

**3.**

Room _____ is between Room 2 and Room 4.

**4.**

Room _____ is after Room 4.

**Week 30 • Making a Map**

# Review

Name _____  Date _____

**Write** and solve the number stories.

1. I have 2 birds. Each bird eats 2 scoops of food a day. How much do they both eat in 1 day?

   _____

2. How much do they eat in 2 days?

   _____

3. I have 12 scoops of food. How much food do I have left after feeding the birds for 2 days?

   _____

**Week 31 • Understanding the Analog Clock**

# It's Hour Time

Name _____ Date _____

**Write** the time that is shown on each clock.

1.  _____ o'clock

2.  _____ o'clock

**Draw** the hour hand on each clock to show the time that is written above it.

3. 11 o'clock

4. 2 o'clock

**Week 31 • Understanding the Analog Clock**

# How Many Hours Have Gone By?

Name _____  Date _____

**Imagine** that each clock started at 12 o'clock. How many hours have gone by?

1. _____ hours have gone by.

2. _____ hours have gone by.

3. _____ hours have gone by.

4. _____ hours have gone by.

96  Level C

# Week 31 • Understanding the Analog Clock

## Review

Name _____   Date _____

**Write** the time that is shown on this clock.

1.    _____ o'clock

**Imagine** that this clock started at 12 o'clock. How many hours have gone by?

2.    _____ hours have gone by.

**Circle** the time that comes after 6 o'clock.

3.  3 o'clock   7 o'clock   2 o'clock   4 o'clock

**Circle** the clock that shows 3:30/Half-past 3.

4.

# Week 32 • The Ten-Dollar Bill and the Dime

## Coins

Name _____  Date _____

### Circle the dime.

1.

### Circle the nickel.

2.

### Circle how many pennies a nickel is worth.

3.

### Circle how many pennies a dime is worth.

4.

# Week 32 • The Ten-Dollar Bill and the Dime

## Bills

Name _____ Date _____

**Circle** the five-dollar bill.

1.

**Circle** the ten-dollar bill.

2.

**Circle** how many dollars a $5 bill is worth.

3.

**Circle** how many dollars a $10 bill is worth.

4.

# Week 32 • The Ten-Dollar Bill and the Dime

## Review

Name _____  Date _____

**Draw** lines to match how many pennies a nickel is worth, and how many pennies a dime is worth.

1.

**Draw** lines to match how many dollar bills a $5 bill is worth, and how many dollar bills a $10 bill is worth.

2.

**Draw** a circle to show how many pennies a dime plus 3 pennies is worth.

3.

100  Level C

# Practice Writing 1–5

**1**

Starting point,
straight down: 1

**2**

Starting point, around right,
slanting left, and straight across right: 2

**3**

Starting point, around right,
in at the middle, and around right: 3

**4**

Starting point, straight down, and straight across right.
Starting point, straight down, and crossing line: 4

**5**

Starting point, straight down, curve around right and up.
Starting point, straight across right: 5

# Practice Writing 6–10

Starting point, slanting left, around the bottom curving up around left and into the curve: 6

Starting point, straight across right, and slanting down left: 7

Starting point, curving left curving down and around right, curving up right and around left to starting point: 8

Starting point, curving around left all the way, and straight down: 9

Starting point, straight down. Starting point, curving left all the way around to starting point: 10